MW01168735

Merry-Go-Rounds

by Art Thomas
pictures by George Overlie

Carolrhoda Books • Minneapolis, Minnesota

The author would like to thank Russell Hehr for his assistance in the preparation of this book.

Manufactured in the United States of America

LIBRARY OF CONGRESS CATALOGING IN PUBLICATION DATA

Thomas, Art, 1952—
Merry-go-rounds.

(Carolrhoda on my own books)
Summary: Presents the history of merry-go-rounds which had their beginning in Arabia more than nine hundred years ago.

1. Merry-go-round—History—Juvenile literature. [1. Merry-go-round—History] I. Overlie, George. II. Title.

GV1860.M4T47 688.7 81-3825
ISBN 0-87614-168-8 AACR2

2 3 4 5 6 7 8 9 10 86 85 84 83

Who doesn't love a merry-go-round?
The music plays.
The horses go round and round,
up and down.
Merry-go-rounds are popular
all over the world.
And they have had a long
and surprising history.

The story of merry-go-rounds
began over 900 years ago.
It all started in the country of Arabia.

The Arabians were great horsemen.
They used horses to work.
They rode horses to hunt.
They rode horses to travel.
They rode horses to war.
They even played games on horses.

One of their favorite games
turned out to be
the beginning of merry-go-rounds.
Men filled clay balls with perfume.

Then they rode their horses
around in a circle.
As they rode, they threw the ball
to one another.

Imagine playing a game of catch
and riding a horse at the same time!
That's not easy!

Only the best horsemen could play.
Usually those men were soldiers.
So the game was called "Little War."

Around the year 1100,
some real wars began.
They were called the Crusades.
Soldiers from Europe came to fight
soldiers in Arabia.
While they were there,
they saw "Little War" being played.
When they went home,
they took the game with them.

By the year 1500,
people were playing "Little War"
in France, Italy, and Spain,
as well as in Arabia.

The Spanish word for “Little War”

is *carosella.*

And to this day

we often call merry-go-rounds *carousels.*

As time went on,
the game became very fancy.
Players decorated their horses.
They dressed up in beautiful costumes.

They even wore wigs.
Musicians played during the games.
Prizes were given to the best riders.
People came from all around to watch.

By 1630, the rules of the game
had changed.
Instead of throwing a ball,
the riders carried spears.
Rings were hung by ribbons
from trees or poles.
The riders rode in circles
around the poles.
They tried to spear the rings.
When they did, the ring and the ribbon
came off the pole.
The ribbons blew merrily in the wind.
What a pretty game to watch!

At first this game was only for noblemen.
But other people wanted to play too.

Soon carousels began to appear at fairs all over Europe.

Then a man in France had a new idea.
His job was to teach young princes
how to ride in the carousels.
In 1680, he invented a machine
to help him.
First he made a couple of horses
out of wood.
Then he hung those horses
from a long bar.
He attached the bar
to a strong pole.
By pushing the bar,
two men could turn the machine.
The princes sat on the horses.
They practiced spearing rings.

The French man's idea caught on.
Over the next 150 years,
people all over the world
were inventing carousel machines.
Some were turned by horses.
Some were turned by mules.
And some were turned by people.
Carousels were popular in Europe.
They were popular in America.
They were popular in Japan.
They were so popular that in 1729,
a man named George Alexander Stevens
mentioned them in a poem.
He called them merry-go-rounds.
And they have been called that ever since.

People all over the world
wanted to ride on merry-go-rounds.
Rich people wanted to put
fancy merry-go-rounds in their parks.

Soon the wooden horses
were getting fancier and fancier.
They were carved to look real.
Then they were beautifully painted.

People wore their best clothes
when they rode on these merry-go-rounds.
What fun they had!

But the ride was not fun
for the animals and men
who ran the machines.
They had to work very hard.

Then, in the 1860s,
another man in France had an idea.
He invented a merry-go-round
that didn't need a man to turn it.
It didn't need a horse or a mule either.
On his merry-go-round,
the horses had pedals.
They were like bicycles.
Everyone on the ride had to pedal.
The harder they pedaled,
the faster the merry-go-round moved.
Now the people who rode
had to do the work.

Then a man in England
decided to use a steam engine
to turn his merry-go-round.
Steam engines had been invented
in the 1700s.
They made all kinds of work easier.
They were used in factories.
They were used on farms.
They were used to run trains.
But they had never been used before
to turn merry-go-rounds.

Steam engines were very powerful.
They were more powerful than many horses.
So merry-go-rounds turned by steam engines
could be much larger.
They could carry more people.

And perhaps nicest of all,
they could play music.
The music came from an organ.
The organ had many pipes.
The steam engine blew air
through the organ pipes.
Each pipe could play only one note.
But each pipe was a different size.
So each one made a different sound.
The music on today's merry-go-rounds
is made in much the same way.

Now merry-go-rounds began to look like
the merry-go-rounds we have today.
There were several rows of horses
on a large platform.

The horses could go up and down.
Roofs were added.
And there were seats for people
who didn't want to ride on horses.

Many of these merry-go-rounds
could travel around.
They were loaded on wagons or trains
and moved from city to city.

The people who built them
tried to outdo one another.
And the merry-go-rounds
of the late 1800s got very fancy.

Some had brass rings.

Riders carried small wooden spears.

They tried to spear the rings.

Some had lions, tigers, rabbits,
and even seals on them
as well as horses.

Others didn't have animals at all.
They had ships or trains instead.
But whatever was on the merry-go-round,
each piece was hand carved.
Then it was carefully painted.

People also decorated
the outsides of merry-go-rounds.
They carved the edges of the roofs.
They carved the centers.
Soon every part of a merry-go-round
was carved and painted.
Merry-go-rounds were very beautiful!

In 1915, merry-go-rounds
were made to run by electricity.
And that is how most of them
are run today.

There aren't many hand-carved
merry-go-rounds anymore.
But merry-go-rounds are still beautiful.
They are still popular all over the world.
And they are still a lot of fun.